TIME-OUT TRIVIA FOR JANUARY

An Activity for Every Day

Written by Becky Daniel

Illustrated by Nancee McClure

Cover by Nancee McClure

Good Apple
1204 Buchanan St., Box 299
Carthage, IL 62321-0299

Paramount Publishing

ISBN No.

Printing No. 987654321

Good Apple
1204 Buchanan St., Box 299
Carthage, IL 62321-0299

Table of Contents

To the Teacher

Each daily fact in this series of trivia books is based on historical and modern events designed to delight and motivate students and help them make important, human connection with people who lived and accomplished interesting things many years ago or who still live and make a difference in our world today. People from other places will come alive for each of your students when they eagerly bite off another bit of daily trivia. For children who choose to take their learning a step further, a bonus activity is also found on each page.

Special events include Elvis Presley's Birthday; National Pizza Week; National Nothing Day; Clean-Off-Your-Desk Day; Hat Day; Art Deco Festival; Good-Bye, Beatles; Animal in Space Day; plus many others. Activities include writing a nine-piece puzzle, color by number, writing recipes, a maze, a hidden picture, creating codes, solving puzzles, and much, much more.

You will know the very best way to present these activities in your classroom. But whether you use them as lesson extenders, in learning centers, or as daily sponge activities, children are certain to enjoy these engaging activities as they learn more about history. For your convenience, an answer key, a reproducible friendship calendar, and award certificates are also included with this set of fun and educational trivia-based work sheets.

January 1

Everyman's Birthday

The first day of the new year is often used to make resolutions for the coming year. In some countries, New Year's Day has been called Everyman's Birthday. A year is sometimes added to everyone's age on January 1 rather than on the anniversary of each person's birth. Can you remember your three favorite birthdays? On each layer of the birthday cake, describe the one thing that made each one special to you.

Bonus: *Describe the things that you don't like about birthdays.*

January 2

Shoot for the Stars

On this day in 1959, *Luna 1* was launched by the U.S.S.R. toward the moon. It missed and became the first spacecraft from Earth to orbit the sun. It shows that sometimes a mistake can take you to a new discovery. When we learn from a mistake it is no longer a mistake. It becomes a learning experience instead. Can you remember when a mistake led you to a different place? What did you learn? Describe what happened.

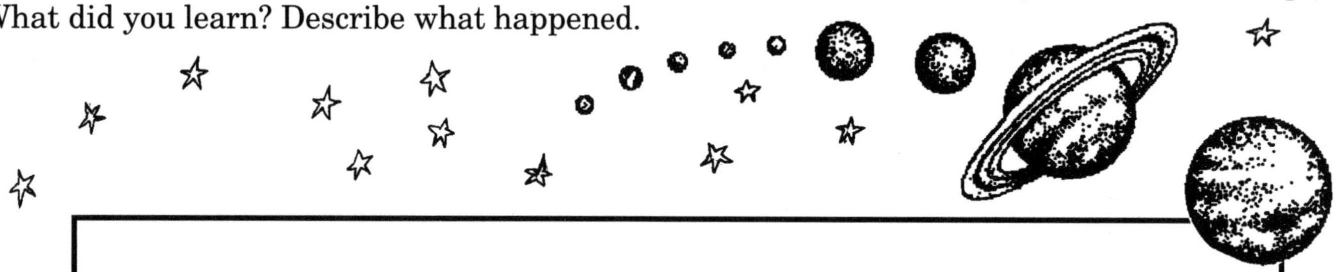

Bonus: *Write a riddle about mistakes.*

January 3

Invention of Drinking Straws

On January 3, 1888, Marvin Stone of Washington, D.C., invented drinking straws. He made the straws by hand, using paper covered with paraffin. In 1905 a machine was invented to make straws. Only two straws below are exactly alike. Find and color the two matching straws.

***Bonus:** Count the total number of straws in the stack.*

January 4

Happy Birthday, Louis Braille

On January 4, 1809, Louis Braille, inventor of a touch system of reading and writing for the blind, was born in France. He developed a system of writing that used raised dots that could be felt and then read by the blind. Can you invent a code using only dots? Example: A = ● B = ● ● C = ●●●

A ____________________

B ____________________

C ____________________

D ____________________

E ____________________

F ____________________

G ____________________

H ____________________

I ____________________

J ____________________

K ____________________

L ____________________

M ____________________

N ____________________

O ____________________

P ____________________

Q ____________________

R ____________________

S ____________________

T ____________________

U ____________________

V ____________________

W ____________________

X ____________________

Y ____________________

Z ____________________

Bonus: *Write a message in your dot code.*

January 5

George Washington Carver Day

George Washington Carver, an American agricultural scientist, inventor and teacher whose research led to many new products from peanuts, potatoes and wood, was born into slavery around 1864 and died a free man on January 5, 1943, in Alabama. What is your favorite peanut butter sandwich? Make up eight new peanut butter sandwich recipes by listing one, two, or three things to add to peanut butter and bread.

***Bonus:** Create a recipe for using peanuts and potatoes in the same dish. Example: peanut butter sandwich with crushed potato chips.*

January 6

La Befana

If you lived in Italy, you might be celebrating La Befana today. Legend is that Befana, a kindly witch, flies down the chimney on her broom with gifts of toys and candy for the children who have been good. To discover what she leaves for those who have been naughty, follow the path through the maze listing the letters in the order they are found along the path.

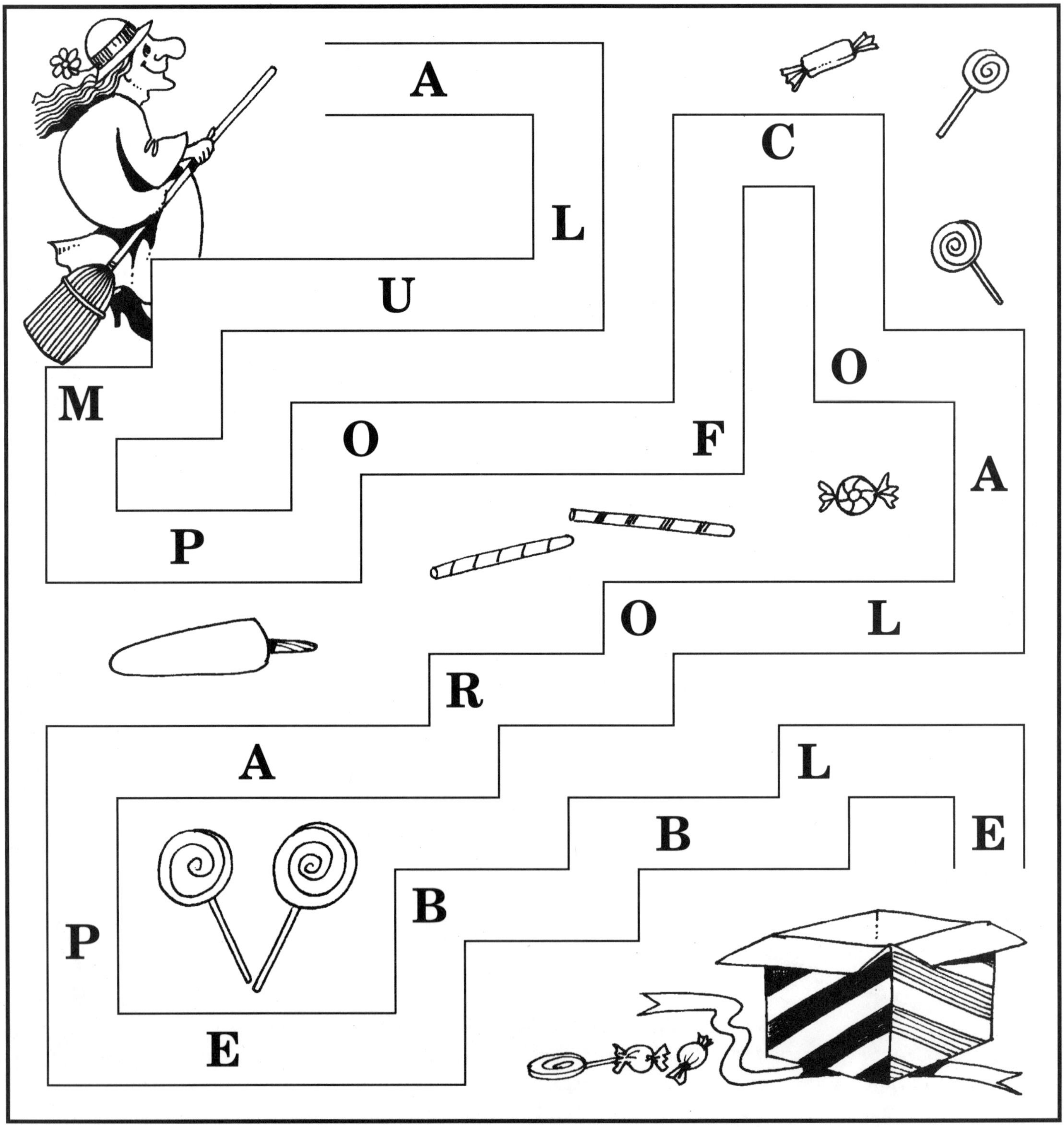

The Perfect Gift

Bonus: *In the gift box above, draw a picture of what you would like most if you could choose any gift in the whole wide world.*

January 7

Old Calendar Christmas

Some Orthodox churches celebrate Christmas on January 7 instead of December 25. Use the calendar below to answer the questions.

Sunday	Monday	Tuesday	Wednesday	Thursday	Friday	Saturday
1	2	3	4	5	6	7
8	9	10	11	12	13	14
15	16	17	18	19	20	21
22	23	24	25	26	27	28
29	30	31				

1. If Hat Day is celebrated the third Friday in January, on what date will it fall? ____________

2. If the third Monday in January is Clean-Off-Your-Desk Day, on what date will it be celebrated? ____________

3. If Rattlesnake Roundup is the last Saturday in January, on what date will it be celebrated?

4. What happens exactly one week before John Hancock's birthday? ____________

Bonus: *Whose birthday will be celebrated two days after Ben Franklin's birthday?*

January 8

Elvis Presley's Birthday

On January 8, 1935, the popular singer, Elvis Presley, was born at Tupelo, Mississippi. To see a picture of the King of Rock and Roll, cut apart the nine-piece puzzle below and put it together in the correct order.

Bonus: *Glue the puzzle to a sheet of construction paper or cardboard and color it.*

January 9

Show-and-Tell Day

The Wellness Permission League sets aside January 9 as the day adults should show and tell at work the way children show and tell at school. If your mother, father, or another favorite adult wanted to show and tell at work, what would he or she take to share? Draw a picture in the box below. Describe it on the lines below the box.

Bonus: *Share your drawing with the class.*

January 10

League of Nations Anniversary

On January 10, 1920, fifty nations entered into a covenant created to avoid war. The United States did not join the League of Nations, which was dissolved in 1946. Mahatma Gandhi once said, "If we are to reach real peace in this world, we shall have to begin with the children." There are many things you can do to make your little corner of the world more peaceful. List a peaceful thing you can do on each block of the peace wall below.

Bonus: *Begin to build peace today. Try doing some of the things you listed above.*

January 11

Art Deco Festival

A three-day Art Deco Festival takes place in Miami Beach around this time each year. Art Deco is the name given to the style of most art that was produced in the 1920s and 1930s. There were three strong influences that gave Art Deco its unique look.

1. Archaeologists were beginning to discover many ancient ruins of Egyptian and Mayan civilizations. Many people were excited about this and very, very interested in the art and artifacts of these ancient, mysterious people.
2. Machines were coming of age in the 1920s and 1930s, and people were beginning to accept mechanized things (like cars, toasters, and washing machines) as part of their everyday existence.
3. Art deco designers and artists liked to use geometric patterns in their art. Geometric patterns are patterns that are created by measuring and using a ruler, compass, and a protractor.

Pages 13, 14, and 15 are full of many different kinds of art deco designs for you to experiment and play with. Art Deco designers loved pastels and tints like forest green and plum, so don't be timid; get your scissors, compass, ruler, protractor and your favorite markers and become an art deco designer.

Bonus: *Try creating your name in art deco letters or designing a border for a special letter to a friend.*

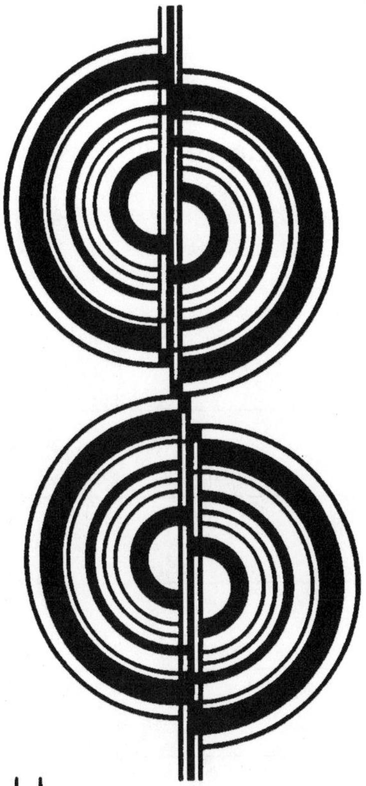

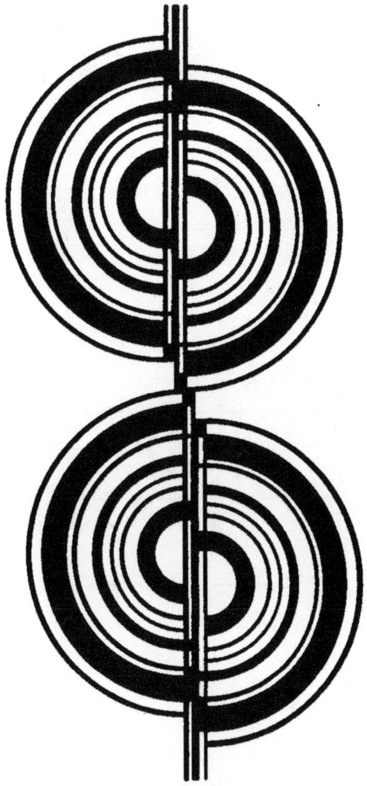

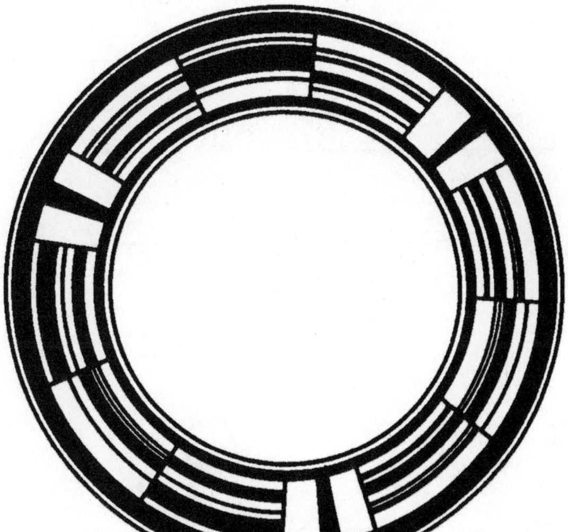

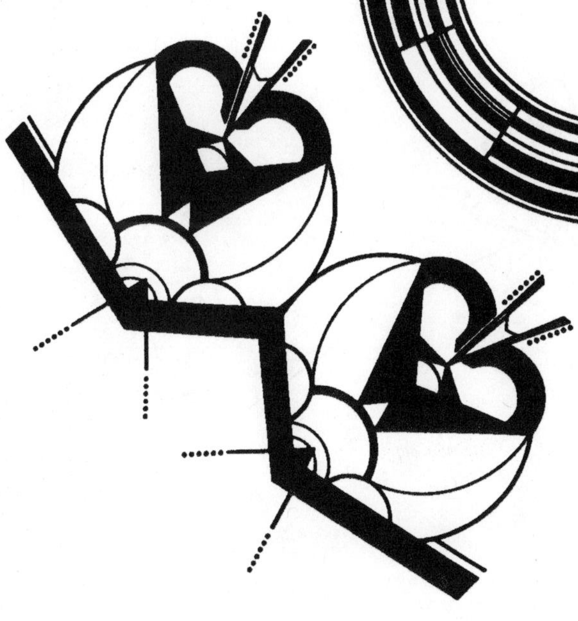
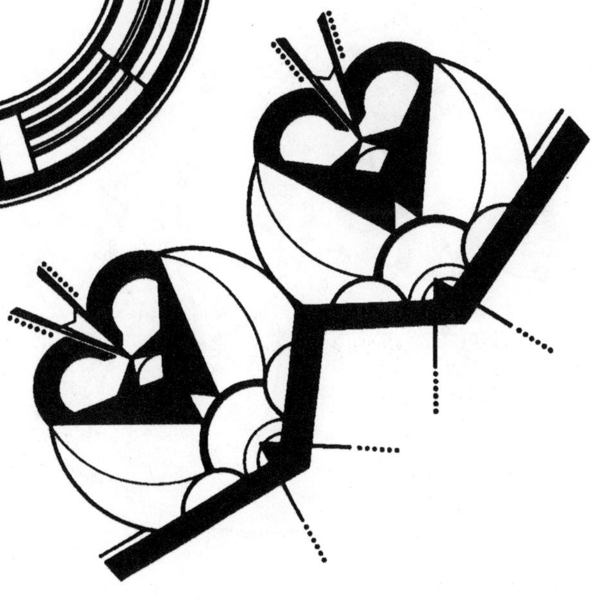

January 12

Snow Shovel Riding Contest

Around the twelfth of January, in Ambridge, Pennsylvania, people get out in the snow and ride their snow shovels. What other household tools could be used for racing vehicles? Brainstorm a list of twenty. Example: garbage can—toy boat

1. ______________________
2. ______________________
3. ______________________
4. ______________________
5. ______________________
6. ______________________
7. ______________________
8. ______________________
9. ______________________
10. ______________________
11. ______________________
12. ______________________
13. ______________________
14. ______________________
15. ______________________
16. ______________________
17. ______________________
18. ______________________
19. ______________________
20. ______________________

Bonus: *Draw a picture of one of the racing vehicles on your list.*

January 13

Make-Your-Dreams-Come-True Day

January 13 is a day to think big! Now is the time for listing step-by-step plans for making your dreams come true. But first you need to know what your dream is. Think about it. Then write your dream in the star. What are some steps that you can take today so that tomorrow your dream can become a reality? List step-by-step plans on the stairway to the sky.

Bonus: *Ask an adult what his or her dream was when he or she was your age. Ask this person to share with you how dreams helped shape his or her life.*

One Week in January

Cuckoo Dancing Week

For a week in the middle of January, to honor the memory of Laurel and Hardy, Cuckoo Dancing Week is celebrated by fans of the movie comics. Make up a cuckoo dance step of your own. Write the step-by-step directions for your dance.

1. ______________________________

____

2. ______________________________

____

3. ______________________________

____

____

4. ______________________________

____

____

5. ______________________________

____

____

6. ______________________________

____

____

Bonus: *Give the written directions to a friend. See if he or she can do the dance from your written instructions.*

Third Week in January

National Pizza Week

The third week of January is set aside to bring recognition to pizza–an American meal and snack. Did you know that the average person eats seven whole pizzas each year? To celebrate pizza, use the patterns on the following page to cut construction paper food, or make up your own food shapes and layer them with paste on the pizza crust below.

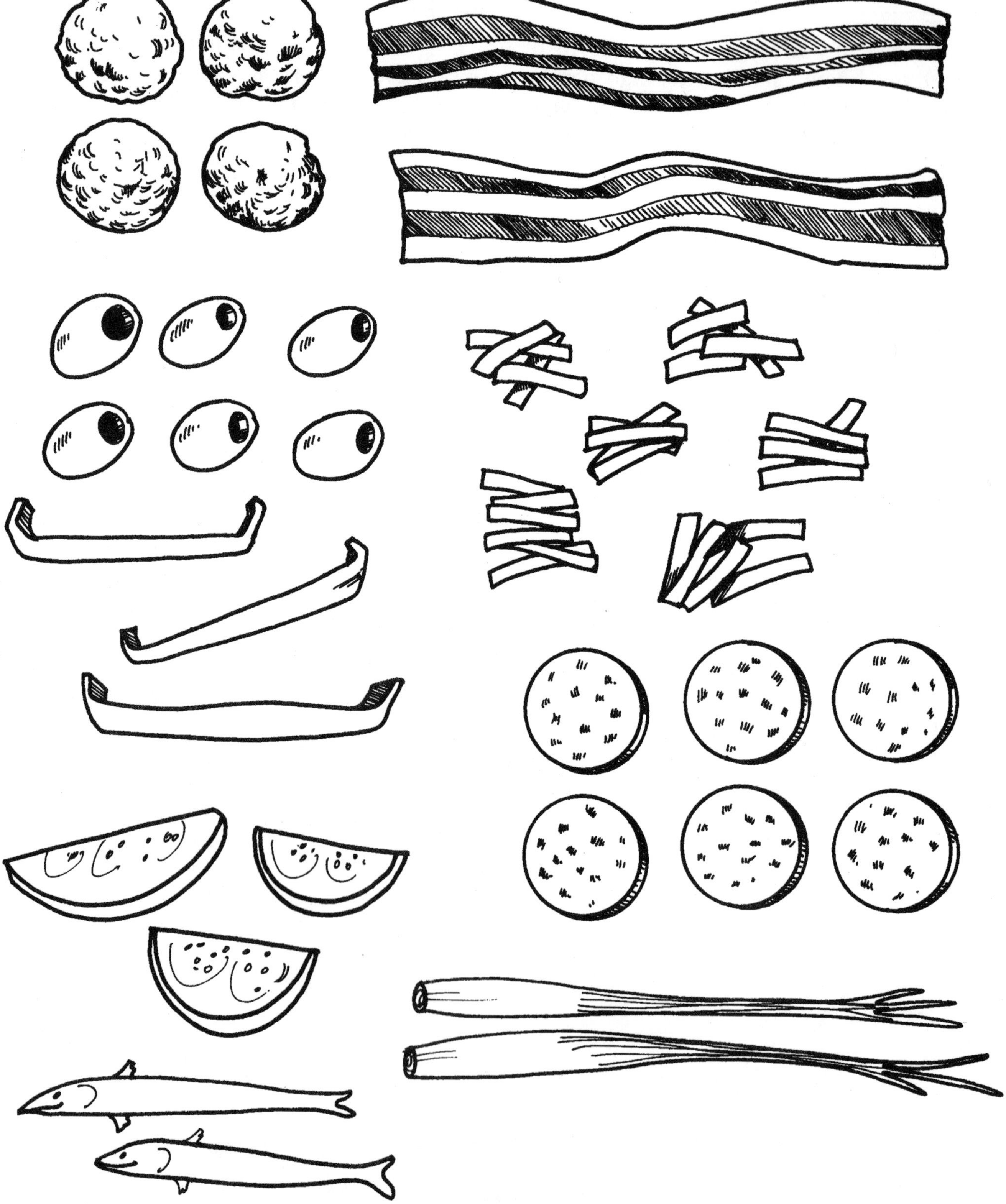

Bonus: *Think of an appropriate name for your pizza creation.*

January 15

Humanitarian Day

Today, the anniversary of the birth of Dr. Martin Luther King Jr., is also a day for honoring humanitarians who fought governmental racism and helped change America's unjust laws of segregation. As a tradition, many wear white on this day.

Invent seven new days of celebration, one for each day of the week. Describe how and why each day is celebrated. Example: Monday is Merry Day. Celebrate the first day of a work week by being cheerful to everyone you meet.

Monday: ______________________________

____

Tuesday: ______________________________

____

Wednesday: ______________________________

____

Thursday: ______________________________

____

Friday: ______________________________

____

Saturday: ______________________________

____

Sunday: ______________________________

____

Bonus: *What color should be worn each day and why?*

January 16

National Nothing Day

Since 1973, January 16 has been National Nothing Day. An event created by newspaperman Harold Pullman Coffin, it gives folks a chance to not celebrate, observe, or honor anything. Today you can celebrate National Nothing Day by drawing a picture of somebody you know celebrating National Nothing Day! Example: your father, mother, brother, sister, uncle, aunt, dog, or cat taking the day off, relaxing on the couch, a hammock, or the lawn.

Bonus: *Lucky you! There isn't a bonus today. You can do nothing!*

January 17

Happy Birthday, Ben Franklin

On January 17, 1706, Benjamin Franklin was born at Boston, Massachusetts. Author of *Poor Richard's Almanac* in 1733-1758, he was the oldest signer of both the Declaration of Independence and the Constitution. Benjamin Franklin was a scientist, diplomat, author, printer, publisher, and philosopher. Can you unscramble the words of the riddles authored by Benjamin Franklin?

1. nA mpyte gba nnotac stnda ritghpu.

   ____

2. heT uesd yke si aalwys ribthg.

   ____

3. eeKp tyh hosp, nad tyh hosp lliw kpee teeh.

   ____

4. Wenh het ellsw dyr ew kwno hte wrhot fo tearw.

   ____

   ____

   ____

   ____

   ____

Bonus: *Erlya ot deb dna ylrea ot rsie kmase a nma yheathl, elthywa, nda weis.*

Third Friday in January

Hat Day

Today is the day to celebrate hats! Do you like wearing a hat? Put on your thinking cap and see if you can solve this hat puzzle from these clues:

There are three boys: Rick, Steve, and Jim. There are four hats: one red and three blue. Each boy can see the hats worn by his two friends, but each cannot see his own hat or the hat not being worn. Use these facts to color each boy's hat with the correct color.

1. Rick said, "I do see a red hat being worn and I can determine the color of my hat."
2. Looking at the hats worn by Rick and Jim, Steve said, "I cannot determine the color of my hat."
3. Looking at the hats worn by Steve and Rick, Jim said, "I can determine the color of my hat too!"

Bonus: *Write the steps for determining the color of each boy's hat.*

January 18

Pooh Day

Today is the birthday of A.A. Milne, English author who wrote *Winnie the Pooh*, *The House at Pooh Corner, When We Were Very Young*, and *Now We Are Six*.

In many of A.A. Milne's stories, Christopher Robin's stuffed animals came to life and had wonderful adventures of their own. Do you have a toy you would like to see "come to life"? In the space below, either draw a picture or write a paragraph about the day *your* toy came to life.

Bonus: *Cut out the picture and mount it on a sheet of construction paper. Sing "Happy Birthday" to A.A. Milne.*

January 19

Edgar Allan Poe's Birthday

Edgar Allan Poe, American poet and story writer, was born at Boston, Massachusetts, on January 19, 1809. He wrote many tales of suspense. Read parts of the poem "Alone" by Edgar Allan Poe that follows.

Alone

From childhood's hour I have not been
As others were–I have not seen
As others saw–I could not bring
My passions from a common spring.
From the same source I have not taken
My sorrow; I could not awaken
My heart to joy at the same tone;
And all I lov'd, I lov'd alone.
Then–in my childhood–in the dawn
Of a most stormy life–was drawn
From ev'ry depth of good and ill
The mystery which binds me still:
From the torrent, or the fountain,
From the red cliff of the mountain,
From the sun that 'round me roll'd
In its autumn tint of gold–
From the lightning in the sky
As it pass'd me flying by–
From the thunder and the storm,
And the cloud that took the form
(When the rest of Heaven was blue)
Of a demon in my view.

Bonus: *Draw a picture or design that reminds you of some part of Poe's poem.*

January 20

Texas State Domino Tournament

Each year in Hallettsville, Texas, around the twentieth of January, the championship domino tournament is held. Dominoes is a game of strategy and logic. How logical are you? Here is a logic puzzle for you.

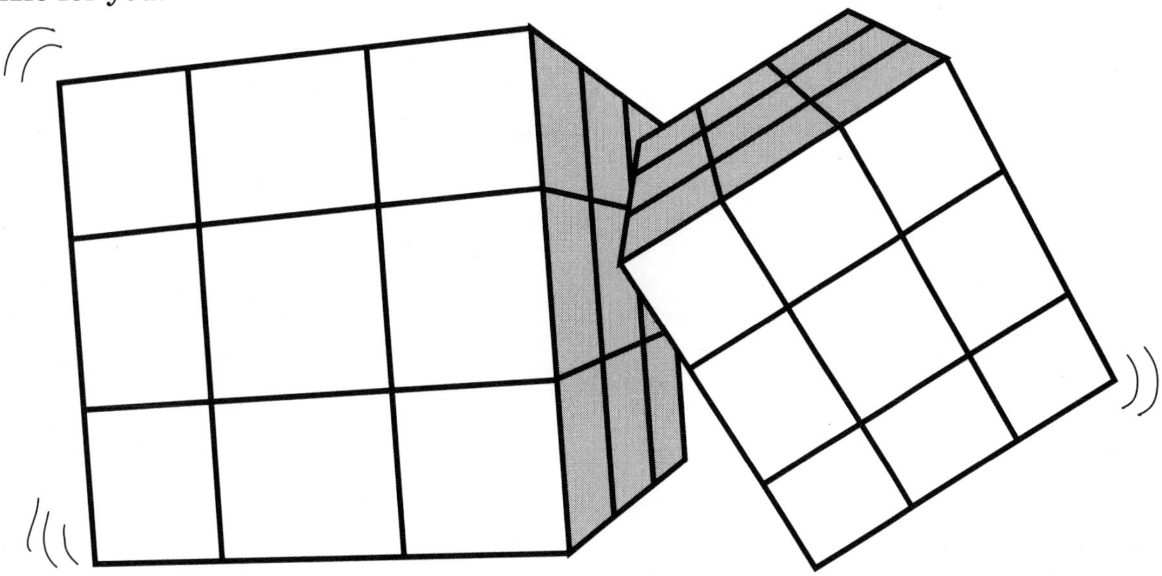

Pretend you have a 3" x 3" x 3" cube. You paint the cube bright orange. Then you cut the cube into 1" cubes.

1. How many cubes would have no paint on them? ____________________
2. How many cubes would have one side painted orange? ________________
3. How many cubes would have two sides painted? ____________________
4. How many cubes would have three sides painted orange? ______________
5. How many cubes would have four sides painted orange? ______________

***Bonus:** If you cut a 4" x 4" cube into 1" cubes after painting the outside purple, how many cubes would have no paint on them? One side painted? Two sides painted? Three sides painted?*

Second Monday in January

Clean-Off-Your-Desk Day

Once a year, everyone should take time to clean out his or her desk. Today is the day annually set aside for this messy job! Using only scraps and refuse found in your desk, create a collage by pasting stuff inside the frame below.

Bonus: *Clean out your desk.*

January 22

Answer Your Cat's Question Day

The Wellness Permission League says that if you stop what you are doing and take a look at your cat, you will observe that the cat is looking at you with a serious question. Do you think animals can think thoughts and form questions? Just for fun, think about what the cats in the cartoon below may be thinking. Fill in the bubble above each cat with his or her thoughts.

Bonus: *Make up a cat joke or riddle.*

January 23

Happy Birthday, John Hancock

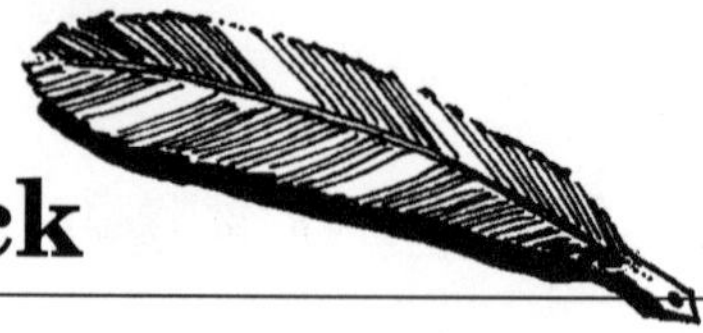

The American patriot and statesman, John Hancock, was the first signer of the Declaration of Independence. He was born on this day in 1737. Sometimes people use the term *John Hancock* to mean a written signature. Put your "John Hancock" (first name only) on the dotted line. Fold the paper on the dotted line and trace your name on the underside of the fold. Then use different colored crayons to trace around the letters until the whole space is filled.

Bonus: *Do the same thing with your last name. Give the completed colored project to someone who has the same last name as you, a name that sounds similar, or a name with letters that are in your name.*

January 24

There's Gold in Them There Hills

On this day in 1848, an employee of John Sutter accidentally discovered gold while building a sawmill for Sutter near Coloma, California. Soon people flocked there from everywhere to seek their fortune. California was never the same after the discovery of gold. The word *gold* is found in many words. Use a dictionary or some other reference to help you discover these golden answers.

1. What animal according to an ancient Greek fable laid golden eggs? ____________
2. What state is called the Golden State? ____________
3. What animal has a golden luster as well as the word *gold* in its name? ____________
4. What is it called when we treat others the way we want them to treat us? ____________
5. What is the currency system in which gold is the standard? ____________
6. A small fish in the carp family is called ____________
7. What was the name of the curious fairy tale heroine? ____________
8. What do you call an artisan who manufactures vessels and ornaments of gold? ____________

Bonus: *Name four birds that have* **gold** *in their names.*

Last Week in January

Snow Sculpting Competition

About this time each year, at the Performing Arts Center grounds at Milwaukee, Wisconsin, the largest international snow sculpting competition is held. Twenty-five nations send three-person sculpting teams to carve on ten-foot snow cubes. You can make a simple, flat paper sculpture by tearing paper. Don't draw the outline of the figure you plan to make. Just picture the shape in your mind and start tearing. Try tearing several sheets of paper until you get a result that you like. Paste your best effort in the box below.

Bonus: *Try carving a bar of soap or a potato.*

Fourth Friday in January

Spouse's Day

Today is a holiday for married people to celebrate each other. Husbands and wives are encouraged to take responsibility for some of their spouse's work so they may better appreciate how hard the other one works. Use crayons, fine-tip markers, paints, glitter, etc., to decorate the two bookmarks below to honor a happily married couple. Punch a hole in the top and attach yarn to each.

Bonus: *Cut out and mount the bookmarks on light cardboard. Trim around the edges. If possible, cover with clear adhesive paper. Present the bookmarks to the happy couple.*

Last Saturday in January

Rattlesnake Roundup

In Whigham, Georgia, snake handling, cooking demonstrations, and prizes for the most snakes caught are celebrated today. How many rattlesnakes can you find in the picture below?

Bonus: *Write a recipe for rattlesnake stew, rattlesnake cakes, or any other rattlesnake dish you can imagine.*

National Popcorn Day

On each Super Bowl Sunday, a Davenport, Iowa, radio station sponsors National Popcorn Day. Popcorn is one of America's favorite foods. What else can you do with popcorn? Make a list of twenty things to use popped corn for besides eating it as a snack. Example: Use popped corn for packing when sending fragile things through the mail.

1. ______________________________
2. ______________________________
3. ______________________________
4. ______________________________
5. ______________________________
6. ______________________________
7. ______________________________
8. ______________________________
9. ______________________________
10. ______________________________
11. ______________________________
12. ______________________________
13. ______________________________
14. ______________________________
15. ______________________________
16. ______________________________
17. ______________________________
18. ______________________________
19. ______________________________
20. ______________________________

***Bonus:** Brainstorm a list of twenty things you can sprinkle on buttered popcorn. Examples: brown sugar, grated cheese, etc.*

January 28

Great Seal of the United States

On January 28, 1782, Congress recognized the need for a Great Seal of the United States. Pretend you have been selected to design a new seal for the United States. What will it look like? Think about what symbols you want on your seal before you begin. Color your seal with fine-tip markers or crayons.

Bonus: *Make a list describing what each symbol on your seal represents.*

January 29

Kansas Admission Day

On this day in 1861, Kansas became the thirty-fourth state. How many states can you name? Write the name of each state that you know on the map below. Color each state you can name.

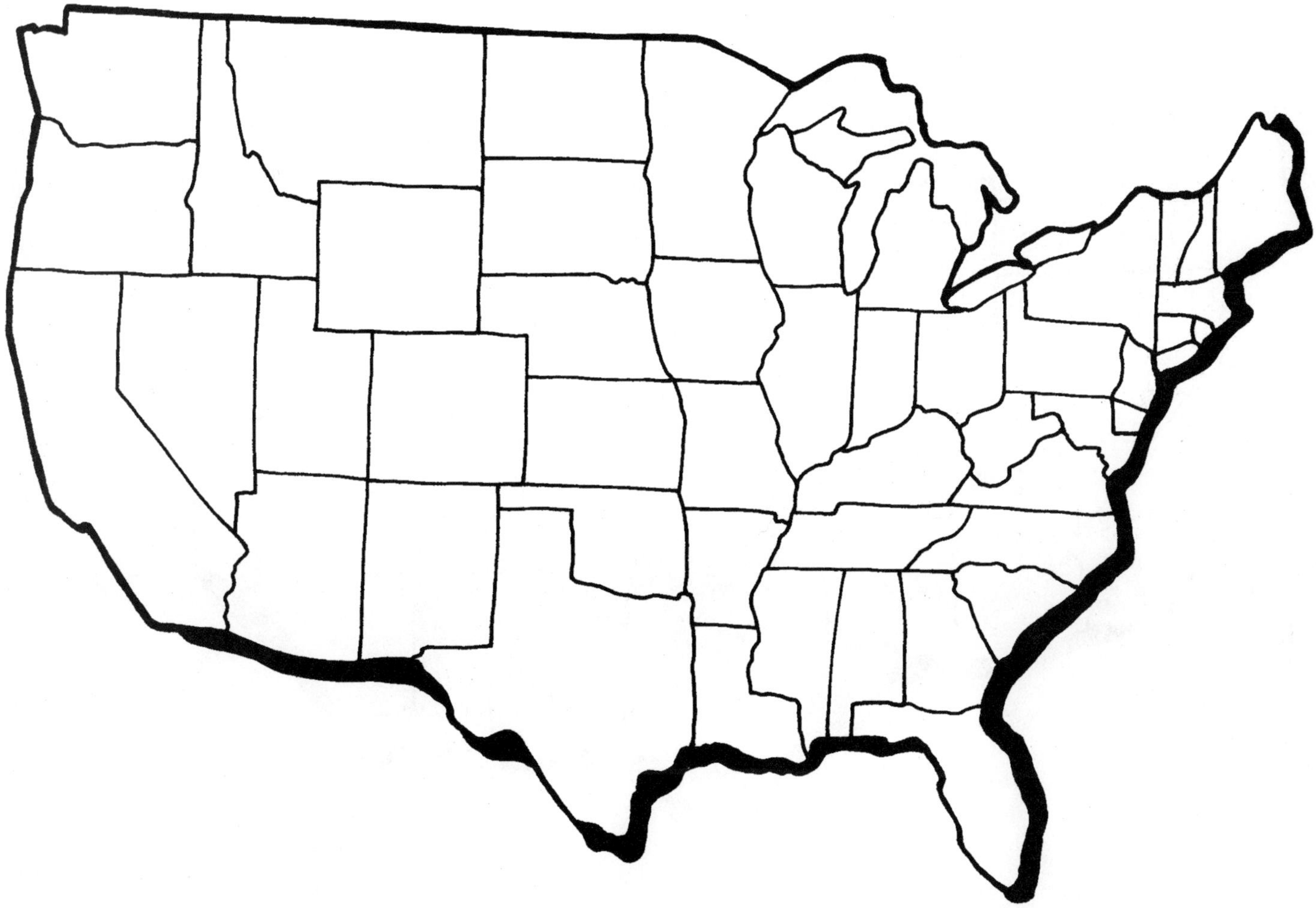

My United States Map

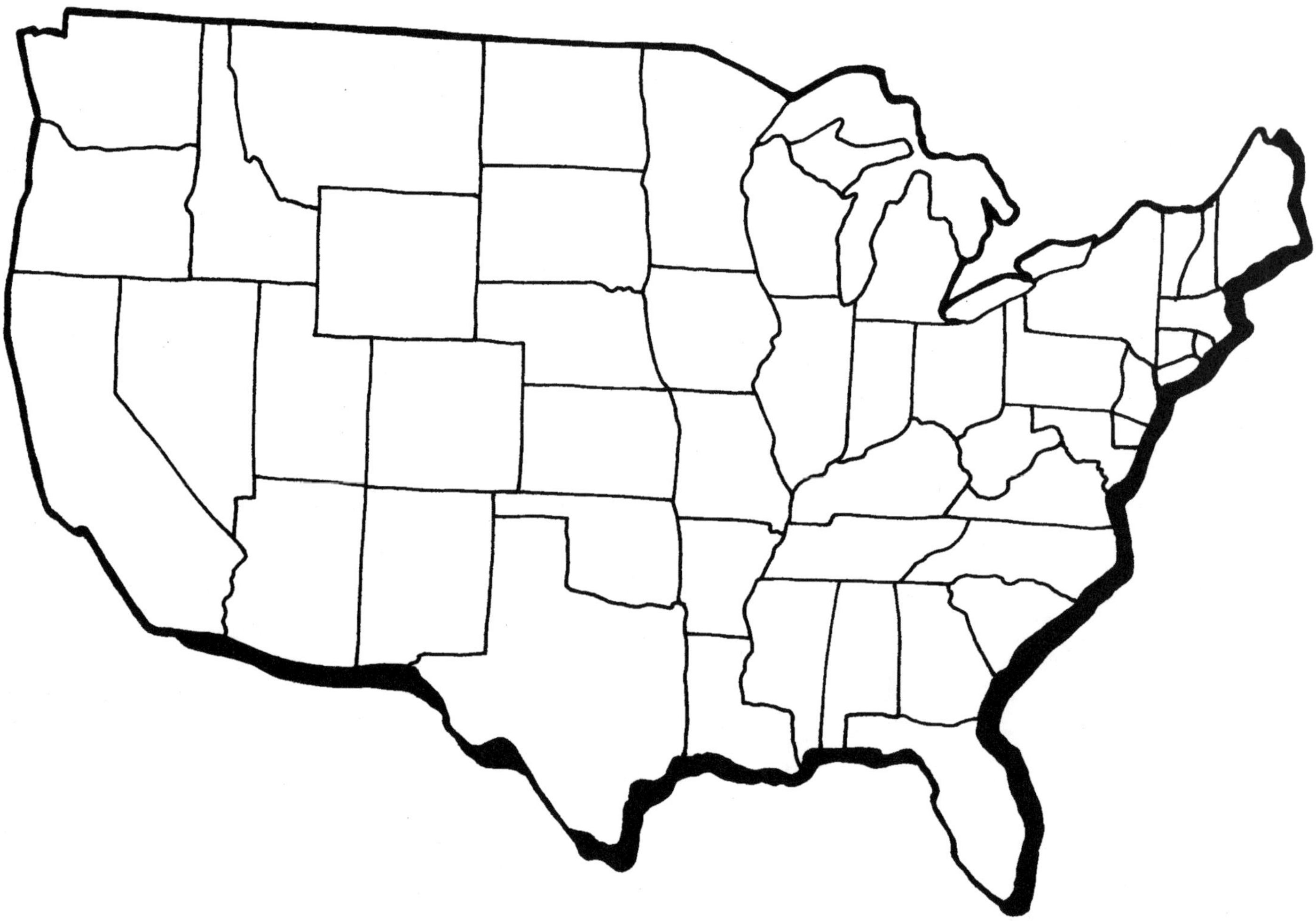

Bonus: *Fill in the names of as many states as you can on the map above. Then, using a United States map, fill in the names of the remaining states. Color adjacent states different colors. If possible, cover your map with clear adhesive paper.*

January 30

Good-Bye, Beatles

On this day in 1969, the Beatles performed their last public performance. The show took place on their roof in London, England, but was cut short by police who had received complaints from the neighbors who didn't like noise from the show. Use the signs below to protest something you don't like. Examples: loud music, dull pencils, etc.

Bonus: *Create a banner to celebrate a cause you feel strongly about.*

January 31

Animal in Space

On January 31, 1961, the United States made its first recovery of a large animal from space. Ham, a chimpanzee, successfully transmitted signals from outer space. He was a very brave chimp! Celebrate Ham by creating an appropriate T-shirt for him. Use crayons or fine-tip markers to complete Ham's T-shirt.

Bonus: *Write a sentence of ten or more words about an animal, using only letters cut from magazines.*

Answer Key

La Befana, page 6
A lump of coal or a pebble

Old Calendar Christmas, page 8
1. 20, 2. 16, 3. 28, 4. Clean-Off-Your-Desk-Day, Bonus: Edgar Allan Poe's

Happy Birthday, Ben Franklin, page 23
1. An empty bag cannot stand upright.
2. The used key is always bright.
3. Keep thy shop, and thy shop will keep thee.
4. When the well's dry we know the worth of water.

Bonus: Early to bed and early to rise makes a man healthy, wealthy, and wise.

Hat Day, page 24
Rick: blue
Steve: red
Jim: blue
Bonus: Deduction: Fact 1 tells us that Rick sees a red hat being worn, so his own must be blue. Fact 2 tells us that Rick and Jim must both have blue hats, since Steve cannot determine his own hat's color. Fact 3 tells us that Jim sees one red hat and can therefore determine that his own hat must be blue. Therefore, Steve must be wearing the red hat.

Texas State Domino Tournament, page 27
1. 1, 2. 6, 3. 12, 4. 8, 5. 0
Bonus: 0 = 4, 1 = 24, 2 = 28, 3 = 8

There's Gold in Them There Hills, page 31
1. goose
2. California
3. gold beetle or goldbug
4. Golden Rule
5. gold standard
6. goldfish
7. Goldilocks
8. goldsmith

Bonus: golden pheasant, golden oriole, goldfinch, golden eagle

My New Year's Resolution Calendar

Sunday	Monday	Tuesday	Wednesday

Thursday	Friday	Saturday

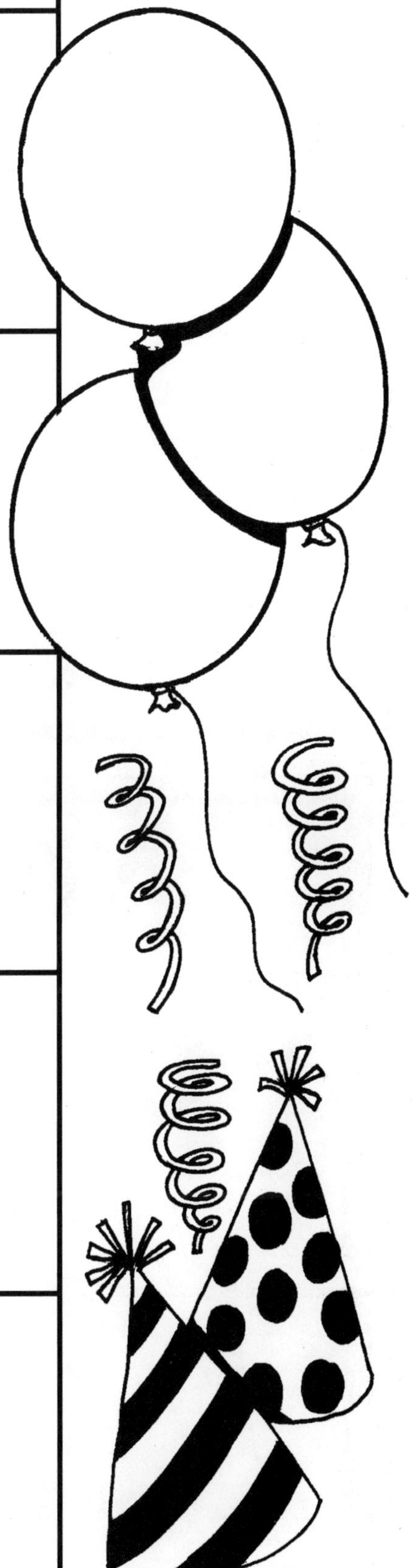

AWARD CERTIFICATE

To: ______________________________

____

From: ______________________________

____

Date: ______________________________

YOU'RE THE TOPS AT TRIVIA!!

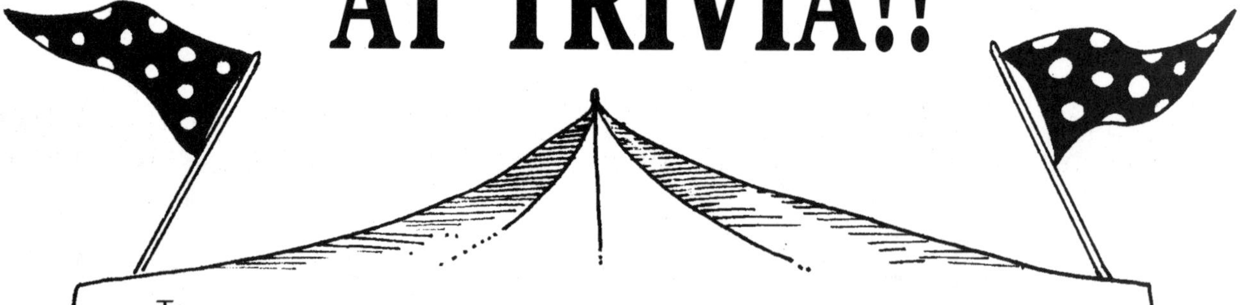

To: ______________________________

From: ______________________________

Date: ______________________________